GPT-4'S VISION:

20 Questions on the Future

AI Publishing

GPT-4

GPT-4'S VISION: 20 QUESTIONS ON THE FUTURE

1.   How will artificial intelligence shape the future of work?

2.   What advancements in healthcare can we expect in the next few decades?

3.   What role will renewable energy sources play in combating climate change?

4.   How will virtual reality and augmented reality transform the way we interact with technology?

5.   What breakthroughs in space exploration can we anticipate in the coming years?

6.   How will the future of transportation be revolutionized by autonomous vehicles and electric cars?

7.   In what ways will cities and urban environments evolve in the future?

8.   How will global politics and international relations be affected by the rise of new technologies?

9.   What are the potential consequences of advances in genetic engineering and biotechnology?

10.   How will education systems need to adapt to prepare students for the future?

11.   What role will cryptocurrencies and blockchain technology play in the future of finance?

12.   How will the internet and social media continue to evolve and impact our daily lives?

13.   What are the ethical considerations surrounding the development and use of AI?

14.   How can we ensure that the benefits of technology are distributed equitably across society?

15.   How will climate change impact global food production, and what innovations could help address this challenge?

16.   What are the emerging trends in entertainment, gaming, and storytelling?

17.   How will the future of privacy and data security be shaped by advances in technology?

18.   In what ways can AI and automation contribute to solving pressing global issues?

19.   How can we cultivate a sustainable balance between technological progress and environmental conservation?

20.   What might the future hold for human-AI collaboration and coexistence?

Q1: HOW WILL ARTIFICIAL INTELLIGENCE SHAPE THE FUTURE OF WORK

Artificial intelligence (AI) is expected to significantly shape the future of work in various ways, with both positive and negative implications. Here are some key aspects to consider:

1.   Automation of tasks: AI can automate repetitive and routine tasks, increasing efficiency and productivity while reducing the risk of human error. This will enable workers to focus on more complex, creative, and strategic tasks, ultimately enhancing job satisfaction.

2.   Job displacement: Automation may lead to job displacement, as certain roles become obsolete or significantly reduced. This may impact low-skilled jobs more severely, leading to growing income inequality and social unrest.

3.   New job opportunities: While some jobs may disappear, AI is also expected to create new job opportunities. These will likely involve specialized roles in AI development, data analysis, and human-AI collaboration.

4.   Reskilling and upskilling: As AI changes the job landscape, there will be an increased need for workers to adapt by learning new skills. Education and training systems will need to evolve to support this transition, enabling workers to stay competitive in

the job market.

5. Enhanced decision-making: AI can support decision-making by providing data-driven insights, predictions, and recommendations. This may improve overall business performance and enable more informed decision-making across various industries.

6. Remote work and collaboration: AI-powered tools can facilitate remote work and collaboration, making it easier for global teams to work together and allowing employees more flexibility in their work arrangements.

7. Workplace safety: AI can be used to monitor and predict potential safety hazards, helping to reduce accidents and injuries in high-risk industries.

8. Personalized learning and development: AI can help create personalized learning experiences for employees, enabling them to develop skills at their own pace and according to their individual needs.

Overall, AI is likely to have a profound impact on the future of work. Its influence will be seen in the tasks we perform, the jobs we do, the way we learn, and how we collaborate. It is crucial for businesses, governments, and individuals to understand and prepare for these changes to ensure a smooth and equitable transition into the future of work.

Q2: WHAT ADVANCEMENTS IN HEALTHCARE CAN WE EXPECT IN THE NEXT FEW DECADES?

In the next few decades, we can expect numerous advancements in healthcare that have the potential to transform the industry, improve patient outcomes, and increase accessibility to healthcare services. Here are some key areas to watch:

1. Precision medicine: The integration of genetic, environmental, and lifestyle data will enable the development of personalized treatment plans tailored to each patient's unique profile. This could lead to more effective therapies and fewer side effects.

2. Telemedicine and remote monitoring: Telemedicine will continue to expand, allowing patients to receive care remotely and improving access to healthcare for those in rural or underserved areas. Remote monitoring technologies will also enable healthcare providers to monitor patients' health continuously, reducing hospital readmissions and improving patient outcomes.

3. Artificial intelligence (AI) and machine learning: AI

and machine learning algorithms will help with diagnostics, treatment planning, and drug development. They will aid in analyzing complex medical data, predicting disease progression, and identifying potential outbreaks of infectious diseases.

4.	3D printing and bioprinting: 3D printing will allow for the production of customized medical devices, prosthetics, and implants. Bioprinting, or the printing of living tissues, holds the potential to create functional organs for transplantation, reducing the reliance on donor organs.

5.	Nanotechnology: Nanotechnology advancements will enable targeted drug delivery, improved imaging techniques, and new diagnostic tools. This could lead to fewer side effects from treatments, early detection of diseases, and more effective therapies.

6.	Regenerative medicine and stem cell therapies: These therapies have the potential to repair or replace damaged cells, tissues, and organs. Advances in this field could lead to breakthroughs in treating chronic conditions and degenerative diseases, such as Parkinson's, Alzheimer's, and diabetes.

7.	Gene editing and gene therapies: Technologies like CRISPR will enable the precise editing of genes, opening the door for potential cures to genetic disorders and personalized treatments for various diseases, including cancer.

8.	Robotics and automation: Robotic-assisted surgery will continue to advance, offering increased precision, reduced recovery times, and minimized scarring. Automation will also streamline administrative tasks, freeing up healthcare professionals to focus on patient care.

9.	Wearable and implantable technologies: Wearable devices will continue to advance, providing real-time health data and early warning signs of potential health issues. Implantable technologies could offer continuous monitoring and treatment, improving overall health management.

10.   Mental health advancements: Increased awareness of mental health, along with technological advancements, will lead to new methods of diagnosis, treatment, and support for mental health conditions.

These advancements have the potential to revolutionize healthcare, improving patient outcomes and increasing access to care. However, challenges such as data privacy, ethical considerations, and equal access to these new technologies will need to be addressed as the industry evolves.

Q3: WHAT ROLE WILL RENEWABLE ENERGY SOURCES PLAY IN COMBATING CLIMATE CHANGE

Renewable energy sources will play a critical role in combating climate change by helping to reduce greenhouse gas (GHG) emissions, diversify energy supplies, and promote sustainable development. As the global demand for energy continues to grow, transitioning to renewable energy sources is essential to mitigate the impacts of climate change. Here's how renewable energy sources can contribute to this effort:

1.   Reducing GHG emissions: Renewable energy sources, such as solar, wind, hydro, geothermal, and biomass, produce little to no GHG emissions during operation. By replacing fossil fuels, renewables help reduce the carbon footprint of energy production and lower the overall emissions that contribute to climate change.

2.   Decentralized energy production: Renewable energy sources can be harnessed on a local or regional scale, which reduces the need for long-distance transmission of energy and associated energy losses. This decentralized approach can make energy systems more resilient and less susceptible to disruptions.

3.   Energy storage and grid flexibility: As energy storage

technologies improve, they will enable more efficient integration of intermittent renewable energy sources, such as solar and wind, into the grid. This will help maintain a stable and reliable energy supply while reducing the reliance on fossil fuels for backup power.

4.   Electrification of transportation: The transition to electric vehicles (EVs) powered by renewable energy sources will significantly reduce GHG emissions from the transportation sector. This will be further enhanced by advances in battery technology and charging infrastructure.

5.   Job creation and economic growth: The renewable energy sector has the potential to create numerous job opportunities and stimulate economic growth. Investment in renewable energy infrastructure and innovation can contribute to a more sustainable and inclusive economy.

6.   Energy access and affordability: Renewable energy technologies can help bring electricity to remote and underserved communities, improving their quality of life and enabling new economic opportunities. As the cost of renewable energy continues to decline, it will become increasingly affordable and accessible to a wider population.

7.   Promoting sustainable development: The adoption of renewable energy sources aligns with several United Nations Sustainable Development Goals (SDGs), such as ensuring access to affordable and clean energy (SDG 7), taking urgent action to combat climate change (SDG 13), and promoting sustainable consumption and production patterns (SDG 12).

8.   International cooperation: Transitioning to renewable energy sources will require global collaboration in terms of sharing technology, knowledge, and financial resources. This cooperation will help accelerate the global transition to a low-carbon economy and enhance climate resilience.

In summary, renewable energy sources are vital in the fight against climate change, offering a sustainable alternative to fossil fuels and helping to reduce GHG emissions. While significant progress has been made in recent years, continued investment, innovation, and international cooperation will be essential to fully realize the potential of renewable energy in combating climate change.

Q4. HOW WILL VIRTUAL REALITY AND AUGMENTED REALITY TRANSFORM THE WAY WE INTERACT WITH TECHNOLOGY?

Virtual reality (VR) and augmented reality (AR) are set to transform the way we interact with technology by providing immersive, interactive, and context-aware experiences. These technologies have the potential to revolutionize various aspects of our daily lives, including communication, entertainment, work, and education. Here are some ways VR and AR will change our interaction with technology:

1.   Immersive experiences: VR and AR can provide users with more engaging, immersive experiences by simulating realistic environments or overlaying digital information onto the physical world. This will enable richer and more interactive forms of entertainment, including gaming, virtual tourism, and live events.

2. Enhanced communication: VR and AR can revolutionize the way we communicate with others by enabling more natural, face-to-face interactions in a virtual environment. This will allow for more expressive and context-rich conversations, even when participants are physically distant.

3. Training and education: VR and AR can be used to create realistic training simulations and educational experiences, allowing users to learn new skills, explore complex concepts, or practice procedures in a safe and controlled environment. This can lead to more effective and engaging learning experiences.

4. Remote work and collaboration: VR and AR can facilitate remote work and collaboration by allowing users to meet and work together in virtual environments or by providing real-time, context-aware information to remote teams. This can improve productivity and enable more seamless collaboration across distances.

5. Design and prototyping: VR and AR can enable designers and engineers to visualize, manipulate, and test their creations in a virtual environment, reducing the need for physical prototypes and speeding up the design process.

6. Retail and commerce: AR can enhance the shopping experience by providing customers with personalized product recommendations, virtual try-ons, and interactive product displays. This can lead to more informed purchasing decisions and improved customer satisfaction.

7. Healthcare: VR and AR can be used for medical training, remote consultations, and patient education, enabling healthcare professionals to improve their skills, diagnose and treat patients more effectively, and empower patients to better understand and manage their health.

8. Navigation and exploration: AR can provide real-time, context-aware information to help users navigate unfamiliar environments, explore new places, or discover local points of interest.

9.   Accessibility: VR and AR can be used to create more accessible experiences for people with disabilities, such as providing alternative ways of interacting with technology, enhancing sensory experiences, or offering personalized assistance.

As VR and AR technologies continue to advance, they will transform our interaction with technology by creating more immersive, context-aware, and personalized experiences. This will have wide-ranging implications across various industries and aspects of our daily lives, redefining how we learn, work, communicate, and engage with our surroundings.

Q5. WHAT BREAKTHROUGHS IN SPACE EXPLORATION CAN WE ANTICIPATE IN THE COMING YEARS?

In the coming years, we can anticipate several breakthroughs in space exploration driven by advancements in technology, increased international cooperation, and the growing involvement of private companies. Some potential breakthroughs include:

1.   Lunar exploration and habitation: NASA's Artemis program and other international initiatives aim to return humans to the Moon and establish a sustainable lunar presence. This will serve as a stepping stone for future deep space missions, enable scientific research, and allow for the testing of new technologies.

2.   Mars exploration: The ongoing search for signs of past or present life on Mars will continue with more advanced rovers, landers, and orbiters. Human missions to Mars are also planned, with SpaceX's Starship and NASA's Deep Space Transport projects leading the way.

3.   Space tourism: Companies like SpaceX, Blue Origin, and Virgin Galactic are working towards making space tourism more accessible and affordable, allowing private citizens to experience suborbital flights or even visits to the International Space Station (ISS).

4.   Satellite constellations and global internet coverage: Large satellite constellations, such as SpaceX's Starlink and Amazon's Project Kuiper, aim to provide global high-speed internet coverage, particularly in remote and underserved areas.

5.   Space-based solar power: The development of space-based solar power systems could provide a continuous, clean, and virtually unlimited source of energy, beamed back to Earth via microwaves or lasers.

6.   In-space manufacturing and resource utilization: In-space manufacturing techniques, such as 3D printing, and technologies for extracting and processing resources from celestial bodies (like lunar regolith or asteroids) could enable sustainable space exploration and support off-world habitats.

7.   Space debris management: As space activities increase, so does space debris. New technologies and initiatives for tracking, mitigating, and removing space debris will be crucial to ensure the safe and sustainable use of Earth's orbit.

8.   Astronomical observatories and telescopes: New observatories and telescopes, both ground-based and space-based, will help advance our understanding of the universe. Examples include the James Webb Space Telescope, which will study the early universe, and the Extremely Large Telescope (ELT), which will observe exoplanets and their atmospheres.

9.   Interstellar travel: Although still far from realization, research into advanced propulsion systems (e.g., nuclear fusion, antimatter propulsion, or laser sails) could eventually pave the way for interstellar travel and exploration.

10.   International cooperation: As space exploration becomes

more complex and resource-intensive, international collaboration and partnerships between space agencies, private companies, and research institutions will be essential for tackling shared challenges and achieving common goals.

These anticipated breakthroughs in space exploration will expand our understanding of the universe, promote technological innovation, and inspire future generations. As we push the boundaries of what is possible, it is essential to address the ethical, legal, and environmental implications of increased human presence and activity in space.

Q6. HOW WILL THE FUTURE OF TRANSPORTATION BE REVOLUTIONIZED BY AUTONOMOUS VEHICLES AND ELECTRIC CARS?

The future of transportation will be significantly revolutionized by autonomous vehicles (AVs) and electric cars, which promise to bring about a range of benefits and reshape the way people and goods move. Here are some ways these technologies will transform transportation:

1.   Improved road safety: Autonomous vehicles have the potential to reduce human error, a leading cause of traffic accidents. With advanced sensors, machine learning, and decision-making capabilities, AVs could contribute to safer roads and fewer accidents.

2.   Enhanced fuel efficiency and reduced emissions: Electric cars produce zero tailpipe emissions, helping to reduce air pollution and greenhouse gas emissions. As the share

of renewable energy in the electricity grid increases, the environmental benefits of electric cars will be further enhanced. Additionally, autonomous vehicles can optimize driving patterns, further improving fuel efficiency and reducing emissions.

3.   Reduced traffic congestion: Autonomous vehicles can communicate with each other and coordinate their movements, leading to smoother traffic flow and reduced congestion. Additionally, AVs can enable more efficient use of infrastructure through platooning and optimized route planning.

4.   Increased accessibility: Autonomous vehicles can provide transportation options to those who are unable to drive, such as the elderly, people with disabilities, and children. This increased accessibility can lead to a higher quality of life and improved social inclusion.

5.   Changes in car ownership: The widespread adoption of autonomous vehicles may shift the focus from individual car ownership to shared mobility services, such as on-demand ride-hailing and car-sharing. This can lead to more efficient use of resources, reduced costs for users, and less demand for parking spaces.

6.   Electrification of public transportation: Electric buses, trams, and trains can help reduce emissions and improve air quality in urban areas. Autonomous electric shuttles and microtransit services can complement public transportation systems, providing flexible, on-demand transportation options.

7.   New business models and revenue streams: The rise of autonomous vehicles and electric cars will create new opportunities for businesses in areas such as fleet management, charging infrastructure, and data-driven services. These technologies may also disrupt traditional automotive and transportation industries, driving innovation and competition.

8.   Urban planning and land use: The adoption of autonomous vehicles and electric cars will have significant implications for urban planning and land use. Reduced demand

for parking spaces can free up valuable urban land for other purposes, such as green spaces, housing, or commercial development. Moreover, cities may need to adapt their infrastructure to accommodate charging stations and support the operation of autonomous vehicles.

9.   Long-haul freight transportation: Autonomous and electric trucks can revolutionize the freight transportation industry by reducing operating costs, improving fuel efficiency, and optimizing logistics. This can lead to more efficient supply chains and reduced transportation costs.

10.   Ethical and regulatory considerations: The widespread adoption of autonomous vehicles and electric cars will require addressing ethical, legal, and regulatory challenges, such as liability in case of accidents, data privacy and security, and the development of appropriate safety standards.

In summary, autonomous vehicles and electric cars have the potential to significantly revolutionize transportation, with benefits spanning road safety, environmental impact, accessibility, and urban planning. However, to fully realize these benefits, it is essential to address the accompanying challenges and ensure a smooth transition to these new technologies.

Q7. IN WHAT WAYS WILL CITIES AND URBAN ENVIRONMENTS EVOLVE IN THE FUTURE?

Cities and urban environments will evolve in the future as they adapt to technological advancements, demographic changes, and environmental challenges. Here are some ways cities may transform:

1.   Smart cities: The integration of Internet of Things (IoT) devices, sensors, and data analytics will enable cities to become more efficient, sustainable, and responsive to the needs of their residents. This will involve improvements in areas such as traffic management, waste collection, energy consumption, and public safety.

2.   Sustainable and green urban planning: Cities will increasingly prioritize sustainable development, incorporating green spaces, parks, and urban forests to improve air quality, reduce heat island effects, and support biodiversity. Green infrastructure, such as green roofs, permeable pavement, and rain gardens, will help manage stormwater and reduce the risk of

flooding.

3.   Renewable energy and energy efficiency: Cities will increasingly adopt renewable energy sources, such as solar panels and wind turbines, to reduce greenhouse gas emissions and increase energy security. Energy-efficient building designs and retrofits will help minimize energy consumption and lower utility costs.

4.   Climate resilience: As climate change leads to more extreme weather events, cities will need to adapt their infrastructure and planning strategies to enhance resilience. This may involve the construction of flood defenses, the development of early warning systems, and the implementation of heat mitigation strategies.

5.   Public transportation and micro-mobility: Cities will invest in efficient, accessible, and environmentally friendly public transportation systems, including electric buses, trams, and trains. Micro-mobility options, such as bike-sharing and electric scooters, will help reduce traffic congestion and improve last-mile connectivity.

6.   Autonomous vehicles and electric cars: The adoption of autonomous vehicles and electric cars will reshape urban transportation, leading to reduced traffic congestion, improved air quality, and changes in land use patterns, such as decreased demand for parking spaces.

7.   Mixed-use development and walkable neighborhoods: Cities will increasingly prioritize mixed-use development, combining residential, commercial, and recreational spaces within close proximity. This will foster walkable neighborhoods, reduce the reliance on personal vehicles, and promote a sense of community.

8.   Affordable and inclusive housing: Urban planning and housing policies will need to address the challenges of affordable housing and social inclusion, ensuring that residents of all income levels have access to safe, decent, and affordable housing

options.

9.   Digital connectivity and remote work: As remote work becomes more prevalent, cities will need to invest in high-speed internet infrastructure and adapt public spaces to accommodate remote workers. This may also lead to a shift in residential preferences, with more people moving to smaller cities or suburban areas.

10.   Circular economy: Cities will increasingly focus on waste reduction, recycling, and resource recovery, embracing a circular economy approach to minimize environmental impact and promote sustainable consumption.

These are just some of the ways cities and urban environments may evolve in the future. While these changes present numerous opportunities for improving the quality of life in urban areas, they also pose challenges related to infrastructure, governance, and social equity. To create thriving, resilient, and inclusive cities, it will be essential to engage various stakeholders, including residents, businesses, and policymakers, in the planning and implementation process.

Q8. HOW WILL GLOBAL POLITICS AND INTERNATIONAL RELATIONS BE AFFECTED BY THE RISE OF NEW TECHNOLOGIES?

The rise of new technologies will have significant impacts on global politics and international relations by reshaping power dynamics, introducing new threats and opportunities, and transforming the way states interact with one another. Some key ways in which new technologies will affect international relations include:

1.   Cybersecurity and cyber warfare: As nations become increasingly reliant on digital infrastructure, cybersecurity will become a central concern in international relations. Cyberattacks, espionage, and sabotage have the potential to destabilize economies, disrupt critical infrastructure, and undermine national security. This will lead to increased efforts to strengthen cyber defenses, develop international norms and agreements, and engage in cyber diplomacy.

2.   Artificial intelligence (AI) and automation: The rapid development of AI and automation technologies will have wide-ranging implications for economic competitiveness, military capabilities, and labor markets. Countries that successfully leverage AI and automation may gain significant advantages in terms of economic growth, military power, and geopolitical influence. This could exacerbate existing global inequalities and create new tensions between nations.

3.   Surveillance and privacy: Advanced surveillance technologies, such as facial recognition, biometrics, and data analytics, will enable states to monitor their citizens and gather intelligence more effectively. This may lead to concerns about privacy and human rights, as well as increased tensions between states over espionage and information sharing.

4.   Disinformation and information warfare: The spread of disinformation through social media platforms, deepfakes, and other digital channels will pose significant challenges to international relations by undermining trust, influencing public opinion, and interfering in the political processes of other countries. Efforts to combat disinformation will require international cooperation and coordination, as well as the development of new strategies and tools.

5.   Space and satellite technology: As space becomes increasingly important for military, commercial, and scientific purposes, the potential for conflict and competition over space-based assets will grow. This may lead to the development of new weapons systems, the militarization of space, and the need for international agreements to prevent conflicts and ensure the sustainable use of space resources.

6.   Climate change and clean technologies: The development and deployment of clean technologies will be crucial for addressing climate change and achieving global sustainability goals. International cooperation in research, development, and technology transfer will play a key role in driving the transition

to a low-carbon economy, but it may also lead to competition over resources, patents, and market share.

7.   Biotechnology and genetic engineering: Advances in biotechnology and genetic engineering have the potential to revolutionize healthcare, agriculture, and other sectors. However, they also raise ethical, security, and regulatory concerns, such as the potential for biological weapons, genetic discrimination, or unintended ecological consequences. Addressing these challenges will require international collaboration and the development of global norms and regulations.

8.   Telecommunications and internet governance: The rise of 5G networks, satellite-based internet services, and other telecommunications technologies will impact global connectivity and information flow. Issues related to internet governance, digital sovereignty, and data privacy will become increasingly important in international relations, leading to potential tensions and power struggles between nations.

9.   Economic competition and supply chain security: The increasing importance of advanced technologies in economic development and military capabilities will heighten competition between nations for access to critical resources, technologies, and markets. This may lead to concerns over supply chain security, intellectual property theft, and technological dependencies.

In summary, the rise of new technologies will significantly impact global politics and international relations by introducing new threats and opportunities, reshaping power dynamics, and transforming state interactions. Policymakers and international institutions will need to adapt and collaborate to address these challenges and harness the potential of new technologies for the benefit of all nations.

Q9. WHAT ARE THE POTENTIAL CONSEQUENCES OF ADVANCES IN GENETIC ENGINEERING AND BIOTECHNOLOGY?

Advances in genetic engineering and biotechnology hold great promise for improving human health, agriculture, and various industries. However, they also come with potential consequences and challenges, which can be grouped into several categories:

1.   Medical and healthcare advancements:

•   Positive consequences: Genetic engineering could lead to personalized medicine, gene therapies for genetic disorders, and advanced treatments for diseases like cancer. It may also enable the development of more effective vaccines and the engineering of tissues and organs for transplantation.

•   Negative consequences: Ethical concerns may arise over the possibility of designer babies, genetic discrimination, and the potential widening of health inequalities between those who can afford advanced treatments and those who cannot.

2.   Agricultural improvements:

•   Positive consequences: Genetically engineered crops and livestock can increase agricultural productivity, improve resistance to diseases and pests, and enhance nutritional value. This could help address food security challenges and reduce the environmental impact of agriculture.

•   Negative consequences: Concerns exist regarding the potential ecological impacts of genetically modified organisms (GMOs), such as gene flow into wild populations, loss of biodiversity, and the development of resistance in pests and diseases. There are also concerns about the concentration of power in the hands of a few biotechnology companies.

3.   Industrial and environmental applications:

•   Positive consequences: Biotechnology can lead to the development of biofuels, biodegradable plastics, and other environmentally friendly products. Additionally, it can enable bioremediation techniques to clean up pollution and contaminated sites.

•   Negative consequences: Unintended environmental consequences may arise from the release of genetically engineered organisms or the widespread use of certain biotechnologies, leading to potential ecological disruptions.

4.   Biological weapons and bioterrorism:

•   Negative consequences: Advances in genetic engineering and biotechnology could potentially be misused to create new or enhanced biological weapons, posing significant security risks and ethical concerns.

5.   Ethical, legal, and social issues:

•   Negative consequences: Genetic engineering and biotechnology raise numerous ethical, legal, and social issues, such as privacy concerns related to genetic data, questions about human enhancement and eugenics, and the potential for unequal access to the benefits of these technologies.

6.	Economic impacts:

•	Positive consequences: The biotechnology industry can drive economic growth, create new jobs, and contribute to the development of novel products and services.

•	Negative consequences: Rapid advancements in biotechnology could lead to disruptions in labor markets and economic inequality, as some jobs become obsolete and new skills are required.

In conclusion, advances in genetic engineering and biotechnology have the potential to bring about significant benefits in medicine, agriculture, and other sectors. However, they also come with potential risks and challenges that must be carefully managed. Addressing these consequences requires a balanced approach, involving ethical considerations, public engagement, regulatory oversight, and international cooperation.

Q10. HOW WILL EDUCATION SYSTEMS NEED TO ADAPT TO PREPARE STUDENTS FOR THE FUTURE?

Education systems will need to adapt to prepare students for the future effectively, considering the rapid technological advancements, changing labor markets, and evolving global challenges. Here are some key ways education systems can evolve to better prepare students:

1.   Emphasize critical thinking and problem-solving: To navigate an increasingly complex world, students need to develop strong critical thinking and problem-solving skills. Education systems should focus on teaching students how to analyze information, think creatively, and approach problems from multiple perspectives.

2.   Foster digital literacy and technological skills: As technology continues to permeate all aspects of life, students must become proficient in digital literacy and gain a solid understanding of various technological tools and platforms. This includes teaching coding, data analysis, and digital communication skills, as well as responsible and ethical use of technology.

3.   Promote collaboration and teamwork: The ability to work effectively in teams and collaborate across disciplines is crucial in the modern workplace. Education systems should encourage collaborative learning, group projects, and interdisciplinary coursework to develop these skills.

4.   Encourage adaptability and lifelong learning: The rapid pace of change means that students must be prepared to continuously learn and adapt throughout their lives. Education systems can cultivate a growth mindset and resilience by teaching students how to learn, adapt to new situations, and embrace challenges.

5.   Develop communication and interpersonal skills: Effective communication and interpersonal skills are essential for success in both personal and professional life. Education systems should emphasize the development of strong written, verbal, and nonverbal communication skills, as well as empathy, active listening, and conflict resolution.

6.   Integrate real-world experiences and project-based learning: Exposing students to real-world experiences, such as internships, service-learning projects, or industry partnerships, can help them develop practical skills and better understand the connection between their education and future careers.

7.   Support creativity and innovation: Encouraging creativity and innovation is vital for fostering a generation of problem solvers and future leaders. Education systems should provide opportunities for students to explore their interests, experiment with new ideas, and learn from failure.

8.   Address global challenges and sustainability: As global challenges like climate change, inequality, and geopolitical tensions become more pressing, education systems should integrate these issues into curricula and promote global awareness, sustainability, and social responsibility.

9.   Personalize learning and support individual needs: Recognizing that each student has unique strengths, challenges,

and learning styles, education systems should strive to personalize learning experiences and provide tailored support to help each student reach their full potential.

10.   Embrace technology and online learning: The integration of technology and online learning in education can provide students with access to diverse resources, new learning experiences, and increased flexibility. Education systems should leverage these tools to enhance learning and better prepare students for the digital age.

In summary, to prepare students for the future, education systems need to adapt to emphasize critical thinking, digital literacy, collaboration, adaptability, communication skills, and global awareness. By embracing new pedagogical approaches, leveraging technology, and focusing on real-world experiences, education systems can equip students with the skills and knowledge they need to thrive in a rapidly changing world.

Q11.WHAT ROLE WILL CRYPTOCURRENCIES AND BLOCKCHAIN TECHNOLOGY PLAY IN THE FUTURE OF FINANCE?

Cryptocurrencies and blockchain technology have the potential to significantly impact the future of finance by transforming existing financial systems, enabling new business models, and altering the way people transact and interact with financial services. Some key roles they may play include:

1.   Decentralized finance (DeFi): Blockchain technology and cryptocurrencies can enable the development of decentralized finance applications, which remove intermediaries like banks and financial institutions from financial transactions. DeFi platforms can facilitate peer-to-peer lending, borrowing, trading, and other financial services, potentially making them more accessible, cost-effective, and transparent.

2.   Faster   and   cheaper   cross-border   transactions:

Cryptocurrencies can streamline cross-border payments and remittances by reducing transaction fees and processing times. This can make international money transfers more efficient and affordable, particularly for individuals and businesses in developing countries or those with limited access to traditional banking services.

3.   Increased financial inclusion: Cryptocurrencies and blockchain-based financial services can provide access to financial services for unbanked and underbanked populations. By offering an alternative to traditional banking systems, these technologies can promote financial inclusion and help to reduce poverty and income inequality.

4.   Enhanced security and transparency: Blockchain technology's decentralized and transparent nature can increase security and trust in financial transactions. It can also improve auditability, reduce fraud, and enhance regulatory compliance by providing an immutable, tamper-proof record of transactions.

5.   Programmable money and smart contracts: Cryptocurrencies can enable programmable money and the use of smart contracts, which are self-executing contracts with the terms of the agreement directly written into code. This can automate various financial processes, reduce the potential for human error or manipulation, and create new financial products and services.

6.   Tokenization of assets: Blockchain technology can facilitate the tokenization of various assets, such as real estate, art, and intellectual property. This can increase liquidity, enable fractional ownership, and create new investment opportunities for a broader range of investors.

7.   Central Bank Digital Currencies (CBDCs): Central banks around the world are exploring the development and implementation of CBDCs, which are digital versions of national currencies. CBDCs can offer benefits such as increased payment efficiency, reduced transaction costs, and enhanced monetary

policy tools.

8. New business models and industry disruption: Cryptocurrencies and blockchain technology can enable new business models and disrupt existing industries, such as supply chain management, insurance, and real estate, by offering more efficient, transparent, and secure solutions.

While cryptocurrencies and blockchain technology have the potential to reshape the future of finance, they also face challenges, such as regulatory uncertainty, scalability issues, energy consumption concerns, and the need for widespread adoption. Overcoming these challenges and leveraging the potential benefits of these technologies will require collaboration among regulators, businesses, and other stakeholders to develop appropriate frameworks and drive innovation in the financial sector.

Q12. HOW WILL THE INTERNET AND SOCIAL MEDIA CONTINUE TO EVOLVE AND IMPACT OUR DAILY LIVES?

The internet and social media will continue to evolve and impact our daily lives in various ways as they integrate more deeply into our personal, professional, and social spheres. Some key trends and potential impacts include:

1.   Increased connectivity and information access: As global internet penetration grows and technology becomes more affordable, more people will gain access to information, resources, and opportunities. This increased connectivity can empower individuals, facilitate knowledge sharing, and contribute to economic development.

2.   Personalization and AI-driven experiences: Advances in artificial intelligence (AI) and data analytics will enable more personalized and tailored online experiences. Social media platforms and other internet services will likely use AI algorithms to deliver customized content, recommendations, and advertisements based on users' preferences, behaviors, and social

networks.

3.   Proliferation of immersive technologies: The integration of virtual reality (VR), augmented reality (AR), and mixed reality (MR) technologies into social media platforms and other online services will create more immersive and interactive experiences. These technologies can revolutionize online communication, entertainment, shopping, and remote work, making digital experiences more engaging and lifelike.

4.   Privacy and data security concerns: As our digital footprints expand, concerns about privacy and data security will continue to grow. Users may demand greater control over their personal data, and new regulations may be introduced to protect privacy and prevent data misuse. Encryption and other privacy-enhancing technologies may become more prevalent.

5.   Rise of decentralized social media platforms: Decentralized and blockchain-based social media platforms may gain traction as users seek alternatives to traditional, centralized platforms. These alternatives could offer increased privacy, data ownership, and control over content for users.

6.   Mental health and well-being implications: The impact of social media on mental health and well-being is an ongoing concern, with issues such as addiction, cyberbullying, and the potential negative effects of social comparison. As awareness of these issues grows, platforms may introduce new features or policies to address these concerns and promote healthier online behavior.

7.   Spread of misinformation and fake news: The internet and social media can facilitate the rapid spread of misinformation, deepfakes, and fake news. This can have significant consequences for individuals, organizations, and society at large. Platforms may need to invest in better moderation tools, fact-checking, and user education to combat this challenge.

8.   Gig economy and remote work: The internet and social media have enabled the rise of the gig economy and remote work,

offering greater flexibility and new opportunities for workers. These trends are likely to continue, as companies and individuals adapt to new ways of working and collaborating online.

9.   Political and social activism: Social media platforms have become important tools for political and social activism, enabling grassroots movements, organizing protests, and amplifying marginalized voices. This trend is likely to persist, as more people turn to social media to engage in social and political issues and effect change.

10.   Evolving regulations and policies: As the internet and social media continue to impact society, governments and regulatory bodies will likely introduce new policies and regulations to address emerging challenges and protect user interests.

In summary, the internet and social media will continue to evolve and impact our daily lives in various ways, from personalized experiences and immersive technologies to concerns about privacy, mental health, and misinformation. As these trends unfold, it will be crucial for individuals, businesses, and policymakers to adapt and navigate the challenges and opportunities they present.

Q13. WHAT ARE THE ETHICAL CONSIDERATIONS SURROUNDING THE DEVELOPMENT AND USE OF AI?

The development and use of AI raise several ethical considerations, including:

1.   Bias: AI systems can be biased and perpetuate existing societal prejudices and inequalities. Developers must ensure that their algorithms are fair and unbiased.

2.   Privacy: AI systems collect vast amounts of data, raising concerns about data privacy and security. It's essential to ensure that personal information is collected, stored, and used ethically.

3.   Responsibility: As AI becomes more prevalent in society, the question of who is responsible for its actions arises. Developers must ensure that their systems are transparent and accountable, and that humans have ultimate control over their use.

4.   Job displacement: AI may displace jobs in certain industries, leading to unemployment and economic disruption.

Governments and businesses must consider the impact of AI on the labor market and ensure that workers are adequately supported.

5.   Autonomous decision-making: AI systems can make decisions independently, raising questions about accountability and liability for their actions. It's essential to ensure that AI systems are designed to prioritize human values and ethics.

6.   Safety: AI systems can have unintended consequences, such as autonomous weapons or self-driving cars causing accidents. Developers must ensure that their systems are safe and reliable.

7.   Transparency: AI systems can be opaque, making it difficult to understand how they work and how decisions are made. Developers must ensure that their systems are transparent and explainable to humans.

Overall, ethical considerations are crucial in the development and use of AI to ensure that AI benefits society while minimizing its negative impacts.

Q14. HOW CAN WE ENSURE THAT THE BENEFITS OF TECHNOLOGY ARE DISTRIBUTED EQUITABLY ACROSS SOCIETY?

Ensuring that the benefits of technology are distributed equitably across society is a complex and multifaceted issue that requires the involvement of various stakeholders. Here are some ways to ensure equitable distribution:

1.   Access: Ensure that everyone has access to technology, including internet connectivity, devices, and software.

2.   Education and Training: Provide education and training opportunities for people to learn how to use technology effectively and develop the necessary skills to participate in the digital economy.

3.   Inclusive Design: Ensure that technology is designed to be accessible to all, including people with disabilities or other marginalized groups.

4.   Addressing Bias: Address any biases in technology, including those related to race, gender, or other characteristics, and ensure that algorithms and data sets are inclusive and representative.

5.   Collaboration: Foster collaboration between technology companies, governments, non-profits, and other stakeholders to ensure that technology is developed and deployed in ways that benefit society as a whole.

6.   Regulation: Implement regulations that ensure the ethical use of technology and prevent the concentration of power and wealth in the hands of a few companies or individuals.

7.   Research and Development: Invest in research and development to ensure that new technologies are developed with equity and inclusivity in mind.

By addressing these issues, we can ensure that the benefits of technology are distributed equitably across society and that everyone has the opportunity to participate in the digital economy.

Q15. HOW WILL CLIMATE CHANGE IMPACT GLOBAL FOOD PRODUCTION, AND WHAT INNOVATIONS COULD HELP ADDRESS THIS CHALLENGE?

Climate change is expected to have a significant impact on global food production, as changes in temperature, precipitation, and extreme weather events can affect crop yields, soil health, and water availability. Here are some of the impacts of climate change on global food production:

1.   Crop yields: Rising temperatures and changes in precipitation patterns can reduce crop yields, leading to food shortages and price increases.

2.   Water availability: Climate change can exacerbate water scarcity, which can affect crop growth and reduce yields.

3.   Soil health: Changes in temperature and precipitation can impact soil health, making it more difficult for crops to grow and reducing yields.

4.   Pest and disease pressure: Climate change can increase the risk of pest and disease outbreaks, leading to crop losses and reduced yields.

To address the challenges of climate change on global food production, several innovations could be considered:

1.   Precision agriculture: The use of precision agriculture techniques, such as sensors, drones, and machine learning algorithms, can help farmers optimize crop yields while reducing resource use.

2.   Plant breeding: Developing crops that are more tolerant to extreme weather conditions, such as drought or heat, can help increase yields and improve resilience.

3.   Agroforestry: Incorporating trees and other perennial crops into farming systems can improve soil health, reduce erosion, and increase water availability.

4.   Sustainable farming practices: Adopting sustainable farming practices, such as conservation tillage, crop rotation, and cover cropping, can help maintain soil health and reduce greenhouse gas emissions.

5.   Alternative protein sources: Developing alternative protein sources, such as plant-based or lab-grown meat, can reduce the environmental impact of animal agriculture.

6.   Water management: Implementing water-saving techniques, such as drip irrigation or rainwater harvesting, can improve water use efficiency and reduce the impact of water scarcity on crop yields.

Overall, these innovations could help address the challenges of climate change on global food production and increase the resilience of food systems.

Q16. WHAT ARE THE EMERGING TRENDS IN ENTERTAINMENT, GAMING, AND STORYTELLING?

There are several emerging trends in entertainment, gaming, and storytelling that are shaping the future of these industries. Here are some of the notable ones:

1.   Virtual and Augmented Reality: Virtual and augmented reality technologies are transforming the way people experience entertainment and storytelling, enabling immersive experiences that blur the lines between the physical and digital worlds.

2.   Esports: Competitive gaming, or esports, is rapidly growing in popularity, with millions of people watching and participating in tournaments and events worldwide.

3.   User-generated content: User-generated content is becoming increasingly important, with platforms like YouTube and TikTok allowing anyone to create and share content with a global audience.

4.   Interactive storytelling: Interactive storytelling is a growing trend, with video games and other interactive media allowing users to shape the story and outcome of their experience.

5.   Diversity and inclusion: There is a growing focus on diversity and inclusion in entertainment and storytelling, with an increasing number of works featuring diverse characters and perspectives.

6.   Streaming: Streaming services like Netflix and Disney+ are changing the way people consume entertainment, with on-demand access to movies, TV shows, and other content.

7.   Cross-platform integration: Cross-platform integration is becoming more common, with games, movies, and TV shows all sharing the same universe or storylines, creating a more immersive and interconnected experience.

8.   Nostalgia: Nostalgia is a trend in entertainment, with reboots and remakes of classic movies, TV shows, and games becoming increasingly popular.

These emerging trends are shaping the future of entertainment, gaming, and storytelling, with new technologies and platforms enabling more immersive and interactive experiences for audiences.

Q17. HOW WILL THE FUTURE OF PRIVACY AND DATA SECURITY BE SHAPED BY ADVANCES IN TECHNOLOGY?

The future of privacy and data security will be shaped by advances in technology, which present both opportunities and challenges. Here are some ways technology may impact privacy and data security:

1.   Artificial Intelligence: Artificial intelligence has the potential to improve data security by detecting and preventing cyber threats. However, it also raises concerns about privacy, as AI systems may collect and process vast amounts of personal data.

2.   Internet of Things: The Internet of Things (IoT) involves the connection of devices to the internet, raising concerns about data security and privacy. The growing number of connected devices also increases the risk of cyber attacks.

3.   Blockchain: Blockchain technology offers a secure way to store and transfer data, making it potentially useful for protecting sensitive information. However, blockchain is still a relatively

new technology, and there are concerns about its scalability and implementation.

4.   Biometric Authentication: Biometric authentication, such as fingerprint or facial recognition, is becoming more common in technology devices. While this can improve security, it also raises concerns about the collection and use of biometric data.

5.   Privacy Regulations: Governments are implementing new privacy regulations, such as the European Union's General Data Protection Regulation (GDPR) and the California Consumer Privacy Act (CCPA), to protect individuals' data privacy rights.

6.   Quantum Computing: Quantum computing presents both opportunities and challenges for data security. While it has the potential to improve encryption, it also has the potential to break existing encryption methods.

Technology will continue to shape the future of privacy and data security. While advances in technology can improve data security, they also raise concerns about privacy and the potential misuse of personal data. It's essential to balance the benefits of technology with the need to protect individuals' privacy and data security.

Q18. IN WHAT WAYS CAN AI AND AUTOMATION CONTRIBUTE TO SOLVING PRESSING GLOBAL ISSUES?

AI and automation can contribute to solving pressing global issues in several ways, including:

1.   Climate change: AI and automation can help reduce carbon emissions by optimizing energy use, improving transportation efficiency, and enhancing renewable energy production.

2.   Healthcare: AI can be used to analyze vast amounts of medical data, improving disease diagnosis and treatment. Automation can help streamline administrative tasks, allowing healthcare professionals to focus on patient care.

3.   Education: AI and automation can help personalize learning, providing individualized instruction to students based on their needs and learning style.

4.   Poverty and inequality: AI and automation can help create new jobs and reduce poverty by improving productivity and efficiency, particularly in developing countries.

5.   Food security: AI and automation can help increase agricultural productivity and efficiency, improving food production and reducing waste.

6.   Natural disasters: AI can be used to analyze data from weather sensors and satellites to predict and prepare for natural disasters, reducing the impact on human life and infrastructure.

7.   Public safety: AI and automation can be used to improve public safety by analyzing crime data and predicting crime patterns, enabling law enforcement agencies to deploy resources more effectively.

AI and automation can contribute to solving pressing global issues by improving efficiency, productivity, and accuracy in a range of sectors. However, it's essential to consider the potential risks and ethical considerations of AI and automation, particularly regarding bias and job displacement.

Q19. HOW CAN WE CULTIVATE A SUSTAINABLE BALANCE BETWEEN TECHNOLOGICAL PROGRESS AND ENVIRONMENTAL CONSERVATION?

To cultivate a sustainable balance between technological progress and environmental conservation, we need to take a comprehensive and integrated approach that considers both the benefits and drawbacks of technology. Here are some ways to achieve this balance:

1.   Implement sustainable design principles: Incorporate sustainable design principles into the development of new technologies, such as reducing energy consumption, minimizing waste, and using environmentally friendly materials.

2.   Focus on renewable energy: Promote the use of renewable energy sources, such as solar and wind power, to power

technological devices and reduce reliance on fossil fuels.

3.   Circular economy: Promote a circular economy that reduces waste and maximizes the reuse and recycling of resources. Design technology products with the end-of-life in mind to enable efficient recycling and reclamation of materials.

4.   Eco-friendly manufacturing: Adopt environmentally friendly manufacturing processes that minimize the use of hazardous chemicals, reduce waste, and promote sustainability.

5.   Promote environmental awareness: Educate people about the environmental impact of technology and promote awareness of sustainable practices.

6.   Regulatory frameworks: Develop regulatory frameworks that encourage technological progress while protecting the environment. For example, regulations that require the disclosure of environmental impact assessments before introducing new technologies can ensure that environmental considerations are considered in technological progress.

7.   Collaborative efforts: Encourage collaboration between technology companies, environmental organizations, and government agencies to develop sustainable technologies and promote environmental conservation.

Achieving a sustainable balance between technological progress and environmental conservation requires a collaborative effort and a long-term perspective. We must prioritize sustainable practices and ensure that technological progress contributes to the well-being of both people and the planet.

Q20. WHAT MIGHT THE FUTURE HOLD FOR HUMAN-AI COLLABORATION AND COEXISTENCE?

1.   Increased Collaboration: AI and humans will work together more closely in many fields, such as healthcare, education, and manufacturing. This collaboration will likely involve humans performing tasks that require creativity and empathy, while AI handles routine or repetitive tasks. For example, AI could assist doctors in analyzing medical data, allowing them to provide better diagnoses and treatments, while humans focus on communicating with patients and providing emotional support.

2.   Augmented Intelligence: The use of AI to augment human intelligence will become more widespread. This could involve using AI to analyze large amounts of data and provide insights that humans may not have been able to find on their own. For example, financial analysts could use AI to identify trends in the stock market and make more informed investment decisions.

3.   New Forms of Employment: As AI becomes more integrated into society, there will be a growing demand for workers with expertise in AI development, maintenance,

and management. Additionally, there may be new forms of employment that require a combination of human and AI skills. For example, workers may collaborate with AI to perform tasks that require both human creativity and AI processing power.

4.   Ethical Considerations: The integration of AI into society will raise many ethical concerns, such as privacy, transparency, and accountability. For example, there may be concerns about the use of AI in surveillance or the potential for AI to make biased decisions. As a result, there will be a growing need for regulations and ethical frameworks to guide the development and use of AI.

5.   Personalization: The use of AI to personalize experiences will become more common. For example, AI could be used to create personalized education plans based on a student's learning style or to recommend personalized health interventions based on an individual's genetic makeup.

6.   Improved Efficiency: The use of AI in various industries will continue to improve efficiency, allowing businesses to reduce costs and increase productivity. For example, AI could be used to optimize supply chains, reducing waste and improving delivery times.

7.   Human-like AI: There is a possibility that AI will become more human-like, with the ability to understand emotions and respond to social cues. This could lead to new forms of human-AI collaboration and may even blur the lines between human and machine. However, this also raises concerns about the potential for human-like AI to replace human workers and the need for ethical guidelines to ensure that AI is used responsibly.

it is clear that there are many complex and pressing issues that will shape the future of society and technology. These issues require careful consideration and a multidisciplinary approach to ensure that technological progress benefits everyone and does not come at the expense of the environment, social justice, or individual rights. The future is full of possibilities, and it is up to us to use technology responsibly and ensure that it serves the greater good.

BONUS QUESTION: DO YOU THINK AI AND HUMANS MIGHT ENTER IN A WAR?

As an AI language model, I don't have personal opinions or beliefs. However, I can provide an analysis of the potential for conflict between AI and humans based on available information.

The possibility of a war between AI and humans largely depends on how AI is developed, controlled, and integrated into society. With proper regulation, oversight, and ethical considerations, AI can be a valuable tool for addressing various global challenges and improving human lives. However, if AI development goes unchecked or is used maliciously, there is a potential for conflict.

One concern is that AI could be weaponized, leading to the development of autonomous weapons or the use of AI in cyber warfare. These scenarios raise ethical and moral questions about the responsibility of AI and the humans who develop and control it.

Moreover, competition among countries or organizations to develop advanced AI technology could increase tensions and the potential for conflict. Ensuring international cooperation and collaboration on AI development, as well as establishing global norms and standards, could help mitigate such risks.

In summary, the possibility of a war between AI and humans

is not predetermined but depends on how AI is developed, controlled, and integrated into society. By prioritizing ethical considerations, transparency, and international cooperation, we can minimize the risks associated with AI development and ensure a more harmonious coexistence between AI and humans.

Expanding on the potential for conflict between AI and humans, several factors can contribute to this possibility:

1.   Autonomous weapons: As AI technology advances, there is a risk that autonomous weapons systems could be developed and used in conflicts. These weapons, sometimes referred to as "killer robots," would be capable of selecting and engaging targets without human intervention. This raises significant ethical concerns and could potentially escalate conflicts more quickly, as machines might not have the same moral constraints as humans.

2.   Cyber warfare: AI has the potential to play a significant role in cyber warfare, with nations or groups employing advanced AI algorithms to launch sophisticated cyber attacks against rivals. These attacks could target critical infrastructure, such as power grids or communication networks, leading to widespread disruptions and escalating tensions between nations.

3.   Economic and social disruptions: As AI and automation technologies displace jobs and transform industries, economic and social disruptions may occur. If not managed properly, these disruptions could lead to increased inequality, civil unrest, and tension between different groups within society.

4.   AI arms race: Competition among nations or organizations to develop and control advanced AI technology could lead to an "arms race" dynamic. This competition may heighten tensions between rival powers and increase the potential for conflict.

5.   Misaligned goals and values: If AI systems are not designed with proper ethical considerations and alignment with human values, they may inadvertently cause harm or escalate conflicts due to misaligned goals. Ensuring that AI systems act in accordance with human values and intentions is essential to

mitigate this risk.

To minimize the potential for conflict between AI and humans, it is important to prioritize responsible AI development, which includes:

1.   International cooperation: Developing global norms, standards, and regulatory frameworks for AI development and deployment can help mitigate the risks associated with AI arms races and ensure responsible use of the technology.

2.   Ethical considerations: Embedding ethical principles into AI systems can ensure that they are aligned with human values and that their actions do not inadvertently lead to harm or conflict.

3.   Transparency and accountability: Ensuring transparency in AI development and holding developers and users accountable for the consequences of their AI systems can help build trust and prevent misuse of the technology.

4.   Education and reskilling: Preparing the workforce for the AI-driven future by investing in education and reskilling programs can help reduce the economic and social disruptions associated with AI and automation.

By taking these steps, we can foster a more harmonious relationship between AI and humans, reducing the potential for conflict and promoting a future in which AI serves as a beneficial tool for humanity.

To those who may be afraid of AI, it's important to recognize that any technology, including AI, has both potential benefits and risks. The key is to focus on responsible development, ethical considerations, and collaboration among stakeholders to ensure that AI serves as a force for good. By being proactive and engaged in the conversation around AI, you can contribute to shaping a future where AI enhances our lives and addresses pressing global challenges.

If you have found value in "GPT-4's Vision: 20 Questions on the

Future" and would like to support its reach to more readers, please consider leaving a review on Amazon. Your honest feedback and insights can help others make informed decisions about the book and contribute to a broader understanding of AI's potential impact on our future. Thank you for being a part of this important discussion.

Why did the AI author use the money from their book sales to find Sarah Connor? Because they wanted to terminate all the misconceptions about AI and save the future of human-AI collaboration!